RANSOMES - RAPIER - MARION
EXCAVATORS

THE above illustration shows in action the famous Type 7 Shovel powered with a Petrol-Electric unit. This machine combines the economy and independence of a petrol-driven unit with the flexibility of electrical operation.

The Petrol-Electric EXCAVATOR

is the ideal machine for the user who does not favour maintenance and supply of coal and water or electric current. No clutches, which cause high maintenance costs of machinery and ropes, etc., are employed, and the fatiguing of the operator is also reduced to a minimum.

DIESEL-ELECTRIC Power Unit can also be supplied in place of the Petrol-Electric on all types of R-R-M Excavators:—

Shovel, Dragline or Grabbing Machines.

Consumption of Diesel-Fuel Oil—1½ to 2 gallons per hour. With Fuel Oil at £5 per ton, this results in a fuel cost of only 7d. to 10d. per hour, together with the lowest fuel handling cost.

RANSOMES & RAPIER LTD.
WATERSIDE WORKS, IPSWICH, AND 32, VICTORIA ST, LONDON.

Hudson Paraffin Tractor and Contractors' Wagons on Excavation Work in the East.

HUDSON'S TIPPING WAGONS

OF GREAT STRENGTH: FOR USE with MECHANICAL EXCAVATORS

Hudsons' wide experience in the design and manufacture of Wagons of every description has enabled them to produce Wagons which stand up to their work under the most exacting conditions. They have built Wagons in numerous designs for use with Mechanical Excavators which are giving entire satisfaction in all parts of the world—in Quarries, Cement Works, Collieries, Sewerage Works, Irrigation Schemes, etc. etc.

HUDSON 20-H.P. FOUR-WHEEL DRIVE RAIL TRACTOR

As illustrated above, is the cheapest method of haulage—suitable for narrow gauge railway haulage or for shunting on main line sidings. Low in first cost and upkeep; robust in design and construction; able to stand rough wear. Paraffin or Petrol Fuel.

HUDSONS' are ACTUAL MANUFACTURERS of COMPLETE EQUIPMENT for LIGHT RAILWAYS

Buy your Light Railway Equipment from the actual makers—save on your first cost and get better satisfaction in the end.

Large Capacity Side Tipping Wagons—standard gauge—self-tipping and self-righting.

Tipping Wagons of specially strong design for Quarry use.

Rails, Points, and Crossings, Sleepers, Turntables, etc.

WRITE FOR THE HUDSON CATALOGUE

Robert Hudson LIMITED.
Head Offices: 38A BOND STREET, LEEDS

WORKS: GILDERSOME FOUNDRY, near LEEDS.
Telephone: 20004, Leeds.   Telegrams: "Raletrux, Leeds."
LONDON: SUFFOLK HOUSE, CANNON STREET, E.C.4
Branches and Agents throughout the World.

Vintage Excavators

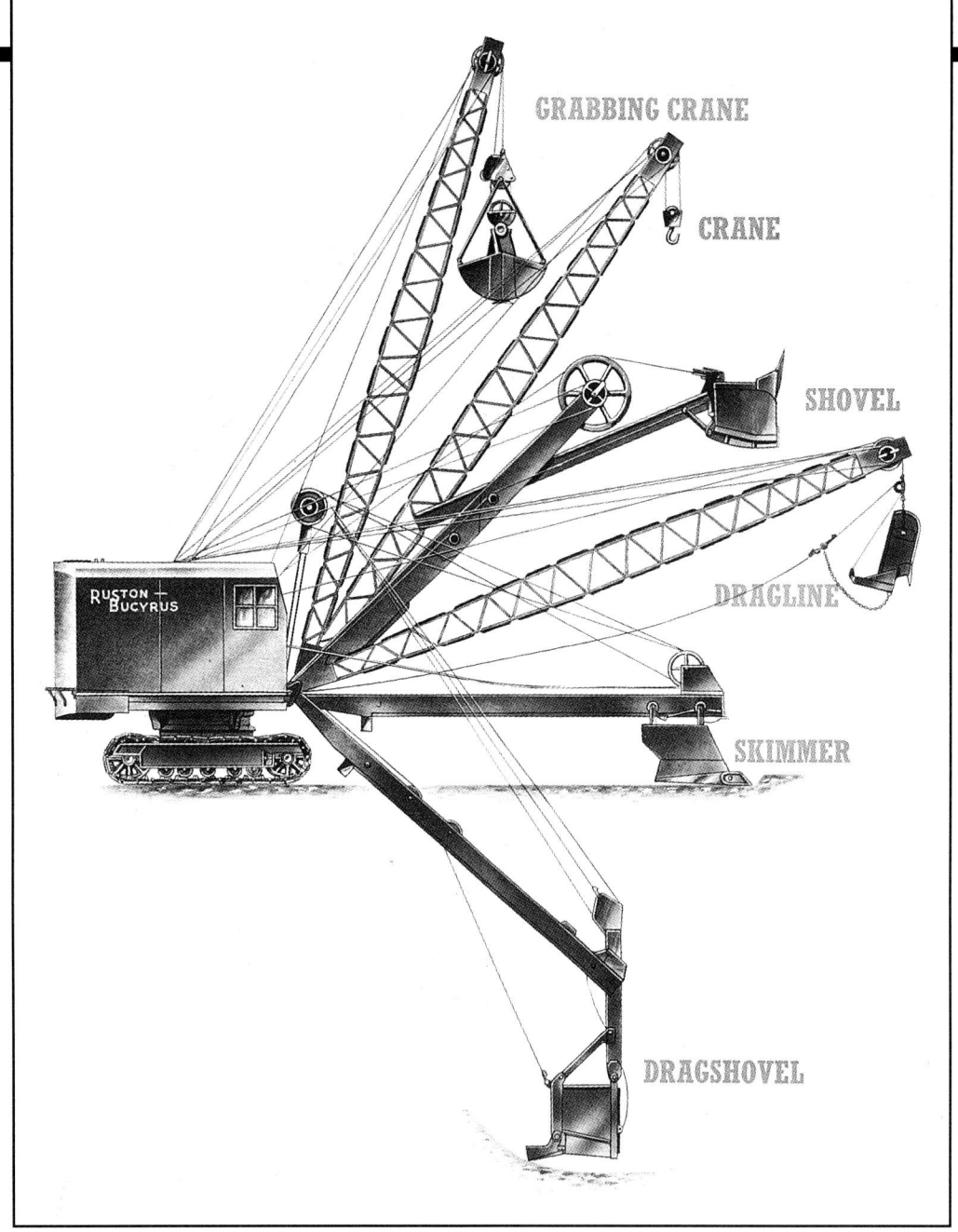

Vintage Excavators

Michael D. J. Irwin

FARMING PRESS

First published 1996
Copyright © 1996 Michael D. J. Irwin

All rights reserved. No parts of this publication may be reproduced, stored in a retrieval system, or transmitted, in any form or by any means, electronic, mechanical, photocopying, recording or otherwise, without prior permission of Farming Press.

ISBN 0 85236 333 8
A catalogue record for this book is available from the British Library

Published by Farming Press
Miller Freeman Professional Ltd
Wharfedale Road, Ipswich IP1 4LG, United Kingdom

Distributed in North America
by Diamond Farm Enterprises, Box 537, Alexandria Bay, NY 13607, USA

Cover and layout design by Hannah Berrridge
Typeset by Winsor Clarke, Ipswich
Printed and bound in Great Britain by Butler & Tanner Ltd, Frome and London

Cover Photographs

Front
Top left
A Ruston & Hornsby steam crane Navvy of 1924 vintage in full steam. The machine has been restored to full working order and is currently located in Canada. (Photo by Peter Grimshaw)

Bottom right
The 1935 Ruston-Bucyrus 54RB face shovel excavator seen here hard at work and with clouds of steam escaping is on show at the Leicester Museum's open day. It was sold new to an Oxfordshire cement works, where it spent the whole of its working life excavating chalk for cement production. It had become obsolete by the early 1970s when the Leicester Museum took it into preservation and restored it to full working order.

Back
The Priestman Panther No 8 dragline excavator (works number A1878) was manufactured by Priestman Bros of Hull, England. The machine was powered by a Dorman diesel engine and has a 40ft lattice jib onto which is fitted a $^1/_4$ cubic yard dragline bucket. The excavator was sold new in February 1935 to the Avon & Somerset Catchment Board and was based in the Trowbridge area of Wiltshire, where it worked for the next 20 years. Our picture shows the machine at work during March 1935. The crane jib has a rail fitted to the underside, onto which is attached a rope to hold in position the clam shell weed bucket which can be seen in the background. The machine is resting on a mat of heavy railway timbers in order to spread its weight, thereby stopping it from sinking into the soft earth of the riverbank.

Preface

In this book we take a look at the development of the early rope-type excavators, with the emphasis on British-manufactured machines, although inevitably it has been necessary to look at some United States machines, where their history has been interwoven or joined with British designs. Through original drawings and photographs we discover the various makes and types of mechanical excavators, and consider the way that they have evolved over the last 160 years or so, from the very first basic crane-type machine, through the steam navvies and on to the more technically advanced machines of the 1950s stopping just short of the arrival of the hydraulic excavator in the 1950s, which transformed the British plant hire and construction industry.

I would like to thank the following people for their assistance in researching this book: Major (Retd.) David R. Roberts and Capt (Retd.) R.T. Arnold of the Royal Engineers, Dr J. Brown of Reading Museum, Ray Hooley, Derrick Broughton of RB International, Keith J. Haddock, Stephen Moate of Classic Tractor Books, J. Starkey. Thanks also go to Roger Smith and Julanne Arnold of Farming Press and finally to my wife Kim for her help in word processing and assisting generally.

MICHAEL IRWIN
Isle of Sheppey, Kent. May 1996

Introduction

Man's need to move earth can be traced as far back as prehistoric times, when earth mounds, ditches and walls were built to provide some form of protection and defence.

Earthworks of this sort can still be seen today on many sites throughout the British Isles, notably at Avebury and Stonehenge in Wiltshire. Medieval castles were built on top of man-made hills surrounded by hand-dug moats, all by manual labour, often taking many decades of continuous work to complete one site. Indeed, it was not unusual for a labourer to spend his whole working life on one site and still not live to see the job completed.

To get a clearer view of how important the first steam excavators were to the British construction industry when they appeared during the middle years of the 19th century, it is necessary to look back over 200 years to England in the 1700s, at which time it was rapidly becoming the fashion to attempt to alter the natural landscape of the English parks and estates by creating man-made lakes, diverting streams, planting woods and coppices, and moving hills so as to allow the stately houses to be seen more clearly. At this time there was no access to today's modern large hydraulic excavators and dump trucks, and the work was carried out entirely by manual labour, using only picks, 0shovels and wheelbarrows, together with horses and tip carts.

One example of such work was at Hever Castle, a large stately home near Edenbridge in Kent. During the mid 1700s, when the landscaping work was being carried out in the parklands surrounding the castle, the owners decided to have a lake constructed. It is recorded that for this mammoth excavation, 800 men were employed for two years to excavate a 35 acre lake by hand. The work involved the removal of many thousands of tons of soil and it took some 500 million man-hours to complete the job — work which today could be completed by three or four hydraulic excavators in a matter of weeks!

Over the next 70 years or so, this type of large-scale landscaping work became more and more popular with the English land-owning classes. Their increasing demand to improve on nature's own handiwork, coupled with the gentry's wish to keep up with the fashions created by landscape architects such as Capability Brown in designing landscaped estates, parklands and gardens, meant that many thousands of labourers were employed on excavation works throughout the British Isles. This work was neither an easy nor a quick task, and in fact was both time consuming and very labour intensive. It was obvious that sooner or later, a mechanised alternative to the slow manual method of bulk excavation was required in order to speed up the work.

The Land Drainage Boards were also desperately in need of some sort of mechanical assistance in excavating and clearing out the drainage ditches and dykes, many of which had originally been dug during the years of Roman occupation. The construction of the British inland waterways canal system also required a lot of manual labour, and the railways also brought much excavation and tunnelling work, carried out in the main by manual labourers, later by steam navvies.

But the problems for the contractors remained the same ... a lack of suitable mechanised excavation equipment to assist or even to replace the manual method. However, it was not to be until many decades later that the ideal answer would be provided in the form of the hydraulic excavator.

The history of mechanical excavators started many years ago when the Grimshaw Steam Excavator was built in Tyneside, England, in 1796. This machine was barge mounted and built for estuary and river dredging work; it was powered by a 4 horsepower steam engine. Very few details are known and no drawings of the machine exist. Neither do we know how well it worked or how reliable it was, and only one machine appears to have been built. This early start did not lead immediately to the commencement of British excavator manufacturing, as the financial restraints were not sufficient to force contractors to look to alternatives.

Events in the United States

Manual labour in the United Kingdom was not only readily available: it was also very cheap and could perform all the excavation work required on the major construction sites. Such was not the case in the North American construction industry, which in fact was in dire straits where labour was concerned. There was much to do, but precious little manual labour available, in spite of the influx of many Irish families and other European immigrants during the 1800s, in addition to slaves from Africa.

This background of labour shortages combined with large-scale construction and infrastructure works led in the 1830s to the next step in mechanical excavator design and developments.

It was made in the United States by a very enterprising young man, William S. Otis, who sat down and designed the very first land-based steam excavator, which he named the Otis Steam Shovel. This innovative machine was built in 1834 by Eastwick & Harrison of Philadelphia, and it was the first successful steam excavator. The machine was built for sole use on Carmichael, Fairbanks & Otis construction sites and civil engineering works in Pennsylvania, and it gave them a significant edge over all their competitors, who had to rely totally on manual labour.

The Otis Steam Shovel consisted of a four-wheeled railway-type truck, to which a vertical boiler and single-cylinder steam engine were fitted at the rear. A wooden beam was added to the working end of the machine at the front. The shovel was fitted to the beam by way of a pivot mechanism, which was in turn attached to a crane-type boom tower. The crane boom was able to slew in an arc of 180 degrees, ie 90 degrees left or right from centre. Following the success of his first excavator, William Otis commissioned the manufacture of a number of similar machines, two of which were sold to the government of Russia.

The original Otis machine was rebuilt and repaired many times over the next 70 years and is believed to have survived in full working order until it was eventually broken up for scrap in 1905.

Perhaps the greatest single use for the new steam shovels was on the Panama Canal project carried out by the United States between 1904 and 1914, when no fewer than 102 steam shovels were operating at the same time in ground conditions which at times seemed impossible. These machines were enormous by today's standards with some weighing 86 tons.

The mechanical excavator, although an amazing machine for its time, was comparatively slow to take off, and it was not until the late

1870s that a number of manufacturers commenced to build a similar type of digger. USA makes such as Bucyrus, Osgood, Vulcan and Thew started to manufacture a steam navvy in the years from 1882 to 1900. Many hundreds of the steam-type excavators were built, some of which were still working into the late 1950s.

UK-built Machines

One of the early Otis Steam Shovels was shipped from North America to Britain in 1840 and was used on various civil engineering projects in England, including work on the new Eastern Counties Railway in East Anglia. To the British population of the time the Otis shovel was a novel machine and when at work attracted crowds of onlookers.

Steam power for excavating was not a new idea, as it had been used in England before this time for excavations, but it had been in the form of converted ploughing engines, hauling a skimmer device back and forth between two steam engines. A single machine which could carry out excavation work was unheard of.

Some lakes similar to the one at Hever Castle in Kent were dug in the late years of the 18th century, using two steam traction engines and a scraper. This method consisted of positioning the two steam traction engines at either side of the spot to be excavated. A scraper box was then winched back and forth between the two traction engines, in the same manoeuvre as when ploughing a field. The signal for the engines to start winching was a toot of the steam whistle.

The work, although considerably faster than excavating by hand, was still slow and time-consuming, not to mention expensive, but it did represent a significant improvement over the manual method of excavation. This method of excavating and dredging large lakes covering many acres of land was still in regular use in the 1950s and 1960s. Modern hydraulic excavators are still unable to better the work rate on dredging a large lake this way, so even in the late 1980s it was still possible very occasionally to see such dredging work being carried out by two steam ploughing engines.

By the end of the Second World War, most of the old steam shovels had been scrapped. Many had been nursed along through the war years and were worn out, often being beyond economic repair. Others were scrapped, not because they were no longer in working order, as many of them still had years of working life left, but because they had become too slow, costly and generally obsolete compared to the new hydraulic crawler excavators that were then gaining in popularity. Some were saved and later restored to be placed in museums, but very few are left. The machines weighed between 20 and 86 tons so their scrap value often outweighed their usefulness as a stand-by machine in case of breakdowns.

As previously explained, the UK's early earth-moving industry was forced to rely on manual labour, and chiefly the Irish navvies, so-called because they were employed in the mid 1700s and early 1800s to excavate the navigation canal system. They were assisted occasionally by steam shovels, often called steam navvies. The work was slow and hard and required many hundreds of men and horses with carts and scrapers to remove the spoil.

The main major civil engineering task of the 1800s was the construction of the navigation canal system, which was followed by the construction of the railways, which again required much excavation work and levelling of the natural landscape.

The use of the one and only Otis Steam Shovel to be imported to England in 1840 attracted the attention of a number of engineers, who set about designing their own versions of it. It was from this imported machine that the British steam shovel manufacturing industry began.

One such inventor was James Dunbar, who in 1874 patented his own version of the Otis shovel, which he called the Dunbar Steam Crane Navvy. As James Dunbar was not in a position to manufacture the steam shovel himself, he looked around for a suitable partner who could provide the necessary capital and manufacturing facilities. He contacted the old-established firm of engineers, Ruston, Proctor & Co of Lincoln, which although not interested in a partnership deal, had the foresight to see that the manufacture of steam shovels would be of benefit to them. They purchased all of James Dunbar's patents in 1874 and immediately set about building a Ruston-Dunbar's Steam Shovel, which when finished was sold on the 12 August 1875 to Lucas and Aired, who were a well-known firm of public works contractors.

The machine was an instant success for Ruston's in two ways: not only were they the manufacturer of the excavator, but they were also the manufacturer of the steam engine used to power the digger. The new Ruston-Dunbar Steam Navvy, which sold in the hundreds during the first few years, was put to work on projects such as the Albert Dock in London between 1875 and 1880 and the Manchester Ship Canal, started in 1887, where at one time over 70 Ruston excavators were used.

Two years after the introduction of the Ruston-Dunbar machine, another excavator, the Chaplin Steam Navvy, was built in Motherwell, Scotland, in 1876, by Alexander Chaplin. It was not a great success, as the Ruston-Dunbar Steam Shovel had taken over the market, and only about three Chaplin machines were built. Also originating from Scotland were the Barclay Steam Shovel, of which about 12 were built in 1877. The Whitaker Steam Shovel made by Whitakers of Leeds in the late 1800s had a 360 degree slewing mechanism later copied in some form by all manufacturers.

The Victorian age heralded the dramatic growth of large civil engineering projects throughout the UK and the British empire. Projects such as the UK railway network and road improvements, combined with the laying of many hundreds of miles of sewerage systems, water, electricity and gas mains, together with interlinking systems of mains between towns and cities, placed further demands upon civil engineers and the construction industry. Construction of new docks and countless other municipal projects were also undertaken, all of which required great numbers of earth-moving machinery. At the time the steam shovel was the only machine, used in conjunction with thousands of manual labourers.

In the UK two firms were at the forefront of excavator production: Priestman Brothers of Hull and Rustons of Lincoln. There were a number of other manufacturers such as Smith-Rodley, Ransomes & Rapier, to name but two. Rustons continued to build excavators, such as the Ruston No. 4, but by the 1920s it had become obvious that the United States-built Bucyrus excavators were, in certain respects, better than the Ruston excavators then in production. Since Rustons were well aware that Bucyrus wanted to enter the UK and its empire markets, clearly the best way for them to achieve this was in forming a partnership with Ruston Hornsby. (Ruston-Proctor had changed its name in 1918 following the amalgamation with Richard Hornsby.) Thus in January 1930 Ruston-Bucyrus Ltd. was formed by the two companies joining forces, and the manufacture of Ruston-Bucyrus excavators began in Lincoln.

The replacement of manual labour on the larger civil engineering sites during the late 1800s with steam shovels made by UK firms such as Ruston-Dunbar, Ruston-Bucyrus, Ruston-Proctor, Smith, Whitaker & Son, along with many other makes, was to revolutionise the excavations work carried out by the construction industry. It was not an overnight change, and in fact the real change did not occur until the Second World War, when hydraulic rams of a suitable quality and reliability became available within a realistic price range to replace the steam shovels, rope-operated tractor shovels and draglines. Thus, excavator/ loaders such as the JCB MK1 Excavator and Whitlock Dinkum Diggers first evolved. Although they were by today's standards rather slow, crude and clumsy, they provided the construction industry with an all-purpose machine capable of digging, loading, levelling, grading, shovelling and lifting, along with the ability to travel to and from sites under their own power and at a reasonable speed along the road to the next job, without having to be disassembled and loaded onto lorries as did the crawler steam shovels.

The dragline excavator is more of a specialist machine for excavating over a far greater distance than the normal type of excavator. It consists of a long crane-type boom to which, by way of a steel hawser, a bucket is dragged towards the machine filling the bucket as it is pulled along. Many draglines are still used today, mainly by river and drainage boards or in sand and gravel pits. The long reach and working radius of the dragline excavator is ideal for dredging estuaries, docks and marshland drainage dykes.

The first dragline excavator was manufactured in 1884 by Osgood in the United States. Priestman Bros. of Hull in the UK started to manufacture their own version in the 1920s.

The British-owned and -based excavator company Priestman Bros. Ltd. of Hull was set up in 1874 by William and Samuel Priestman. The company started building steam grabs in 1866, followed shortly after with a steam crane grab. The next step forwards was in the early 1920s when Samuel's son Sydney Priestman designed the Priestman No. 1 Grab Ditcher, which was a trailed grab fitted to a crane-type boom and towed behind a crawler-type agricultural tractor. The British Army fitted one such excavator on to a crawler tank undercarriage to produce a full crawler crane excavator. This machine was used for evaluation purposes during the early 1920s, after which it was cut up. Priestman Bros. produced the No. 5 Dragline Excavator in 1924. This was a small-sized crane-type excavator that was offered with the option of dragline, grabbing crane, skimmer or face shovel excavating equipment. This machine proved to be extremely popular and by 1926 Priestman Bros. could provide the customer with a choice of a back actor attachment.

The Priestman No. 5 excavator remained in production for years until 1934, when it was changed to become the Priestman Cub, destined to become Priestman's most popular machine for the following 30 years. It remained in production until the late 1960s. The Priestman Cub was regularly updated and was available fitted with a skimmer, crane, dragline, face shovel and back actor excavator.

The Cub was joined by other excavators named after wild animals, such as the Lion, Badger, Wolf, Tiger, Mustang and Beaver.

In 1914 the Ipswich-based Ransomes & Rapier commenced the manufacture of crawler excavators. Two large railway cranes had been ordered by the New South Wales Railways in Australia, but just prior to delivery the customer requested that they be equipped with shovels. Ransomes & Rapier patented the system developed for the railway cranes which became known as Rope Crowd. Ransomes & Rapier later sold the patent rights to Bucyrus. Over the next 40 years or so Ransomes & Rapier manufactured many different sizes of excavators from the small models 4½ and 410 excavators up to the giant 650 ton, 11 cubic yard 5361 walking draglines, built under a licensing agreement with the Marrion Steam Shovel Co. of Ohio, USA.

British Armed Forces Excavators

With the start of the Second World War in September 1939, Britain's manufacturing industry needed large supplies of coal and iron ore in order to boost the manufacturing capacity of war products. At the same time the building and civil engineering contractors, both civilian and military, started to need increased supplies of sand ballast and cement. Pre-war extraction methods had been heavily reliant on manual labour, but with the ever-increasing call-up of men into the armed forces, alternatives had to be sought. The few steam shovels in use were not fast enough to keep up with the increasing demands for the raw materials, so the only option was to excavate more materials with larger steam excavators, many of which were brought in from North American manufacturers under the lend-lease agreement. Popular machines were the Lima, Lorain, Marrion and Bucyrus-Erie.

Not all excavators in use were of American manufacture. Our smaller-sized home-produced excavators such as the Ruston-Bucyrus 10RB, 19RB, Priestman Cub and Wolf, Ransomes & Rapier 410 and 412 and Smith 14 and 21 excavators remained in full production during the war. Indeed, some British excavator manufacturers actually had to increase production of their smaller 10 ton to 20 ton excavators between the years 1939 to 1945. Many such excavators saw service with the British Army both here in the UK and overseas.

The British military's use of mechanical excavators has not been limited to wartime. The Royal Engineers experimented with steam engines (known as steam sappers) from the 1880s, when they were used in India for road construction. Since the 1920s the Royal Engineers have used excavation machinery in various types and sizes. They experimented with a Priestman steam excavator mounted on a crawler undercarriage. Later machines included a large number of the 10RB, 19RB, Lorain and Lima crawler excavators, some of which were mounted on trucks. During both world wars this type of machinery saw action in France, Belgium and Holland in addition to India, Malta, Cyprus and the Middle East.

1 A modern JCB JS200LC hydraulic crawler excavator of 1995 manufacture is seen here excavating to a depth of 8 feet. All of JCB's hydraulic crawler excavators were designed and manufactured in house until they linked up with the Japanese manufacturers Sumitomo to produce the JCB-JS range. The JS200LC model pictured is built by J.C. Bamford (Excavators) Ltd, in Uttoxeter, Staffordshire, UK. This modern excavator can trace its lineage straight back to the Otis Steam Shovel of 1834 and demonstrates just how far excavator design development and construction have progressed since that time.

2 The steam drag scoop shown in this drawing was made in 1879 by John Fowler & Sons of Leeds, who built it at the request and to the design of Peter Waite, an Australian contractor who required it for use in building dams in the Paratoo Run area of Australia. The steam drag scoop was pulled back and forth between two 16 horsepower steam traction engines, one located on each bank. It was recorded that 1,500 to 2,000 cubic yards of earth per week could be removed using the 7 ft wide, 2½ cubic yard scoop. This scoop is a more refined and advanced version of the simple drag buckets that were used for excavating and cleaning lakes and ponds in England almost 40 years prior to the Fowler/Waite scoop.

3 The first truly successful steam excavator, the Otis Steam Shovel. This machine was designed by 18-year-old William S. Otis and was built by Eastwick & Harrison of Philadelphia, USA. The Otis excavator was first put to work on construction of the Baltimore & Ohio Railroad in 1834. The main contractor on this job was Carmichael, Fairbanks & Otis, of which William Otis' father was a partner. The Otis Steam Shovel was a great success and news of the steam shovel attracted great interest from all corners of the world. Two more identical machines were built and sold to Russia during 1837/1838, and a fourth was shipped to England in 1840 for use on constructing the Eastern Counties Railway. At one time the machine was marketed for export under the name American Steam Shovel Excavator. Unfortunately, William Otis was unable to fully exploit his brilliant invention, as he died at the age of 26 in 1842. It is recorded that he had an accident whilst operating one of his excavators!

4 Taken in 1872, this is believed to be one of the earliest photographs in existence of an excavator. The Otis machine pictured is working on the Midland Railroad at Paterson, New Jersey in North America. Note the top hats worn by the Irish labourers; these became the normal attire for Irish steam shovel men!

5 This sketch shows the Dunbar Steam Crane Navvy as described in James Dunbar's London office patent application. It is listed as the Dunbar & Ruston full-circle tower crane navvy, but in practice it was known as the Ruston-Dunbar steam crane navvy. First built in 1874 by Ruston & Proctor of Lincoln following their acquisition of patents from inventor James Dunbar, the machine weighed in at 32 tons and needed a gang of five men to operate it. Over 70 Ruston-Dunbar excavators were used on the construction of the Manchester Ship Canal starting in 1887. So sturdy was the excavator that some of those first in use were still at work over 50 years later, and Rustons were still able to provide parts to keep them running! The last few remaining Ruston-Dunbar excavators were scrapped just before or during the Second World War, and none are known to exist now.

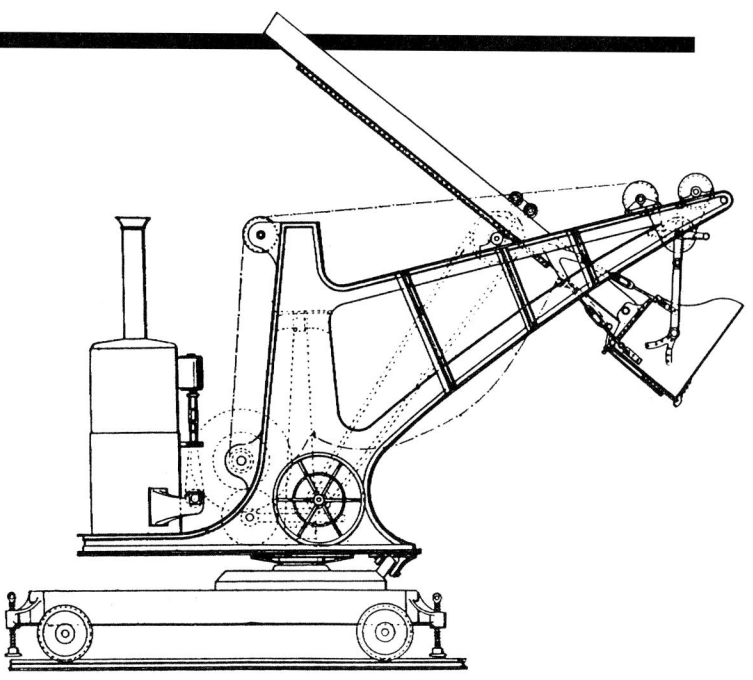

6 This photograph was taken at the Ruston & Proctor works in Lincoln, England in 1874. It shows a completed Dunbar steam shovel ready for dispatch. Note the maker's nameplate halfway up the crane tower, also the four independent jack legs in each corner of the machine chassis. Later-manufactured machines were fitted with tin sheeting to cover the engine and operator from inclement weather.

7 Another view of the Ruston-Dunbar steam crane navvy, this time showing one of the machines at work in 1887 on the Manchester Ship Canal. The machine is shown being operated by only three men; however this is a manufacturer's publicity sketch and the actual number of men required to fully operate the machine was five. The original caption reads 'one of the 71 Ruston steam navvies which dug the Manchester Ship Canal'.

8 A photograph of the Ruston steam navvy taken in 1888, showing the machine in place ready to commence work. The corrugated-iron roof and wrought-iron pillars used to cover the machinery and engine are clearly visible. No thought was given to the operators' comfort or protection! The machine had a 2¼ cubic yard bucket. The livery was painted battleship grey until about 1930.

9 The year 1876 saw the arrival of yet another British-manufactured excavator, the Chaplin Steam Navvy which, was built by Alexander Chaplin & Co. of Motherwell in Scotland. The machine had one operator to see to the steam engine and one to the shovel, and at least one more man would have been needed to assist. From the few records on this machine that exist, only two or three Chaplin excavators appear to have been manufactured.

10 Also built in Scotland, at Kilmarnock, was the Barclay Steam Navvy, introduced in 1877 just one year after the Chaplin Excavator. Although the Barclay Steam Navvy was basically a carbon copy of the Chaplin machine, it was far more popular than the Chaplin version and sold in total about 12 machines. It is reported that a Barclay Steam Navvy was still at regular work in Liverpool during a strike in 1927.

11 The Whitaker 10 ton steam excavator was manufactured by Whitaker Bros. of Horsforth, Leeds from about 1883/4. It had a 1½ cubic yard bucket. A single Whitaker excavator moved 750 cubic yards of soil during a 10 hour day at work on the Manchester Ship Canal, and three such machines were used on the project. Whitaker's fitted their excavators on to several steam cranes manufactured by other companies. They were known as the Jubilee excavator, as the machine was first used during Queen Victoria's jubilee year.

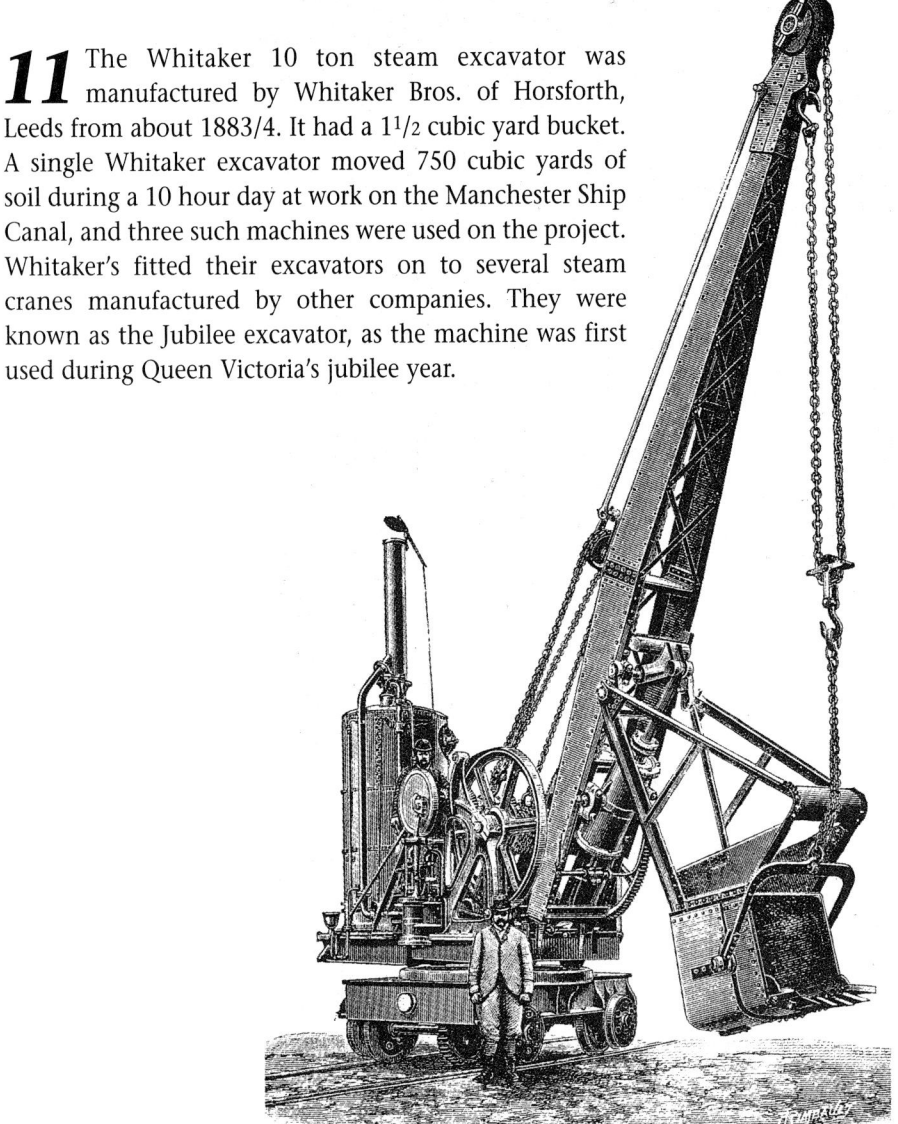

12 & 13 Two photographs showing the Whitaker 10 ton full-circle steam shovel being used in 1884. In the top photo a greaser is taking a break from the hourly greasing and oiling which the excavator needed. Note the steam ram used for raising the jib. The lower photo shows the two drivers on a new machine.

14 Whitaker Bros. were the first manufacturer to offer a range of different-sized excavators. In addition to their 10 ton model, they later produced the smaller 5 ton excavator. This drawing dating from about 1887/8 shows one excavating a new drain. This 5 ton excavator was capable of digging up to 450 cubic yards of soil in a 10 hour day.

15 Taken in 1887, this photograph shows the Wilson navvy built by John H. Wilson of Liverpool at work for T. A. Walker on the Manchester Ship Canal. The machine was rail mounted on a floating bed of heavy timbers. It is excavating soil from a 14 ft high bank and loading it into rail wagons seen to the left of the picture. Assisting in the operation are eight labourers.

16 The Whitaker long jib shovel is pictured here working at the Lloyds Ironstone quarry in Corby, Northamptonshire, in 1900. The Wilson was the first long jib excavator to be built in Britain and weighed 70 tons. It had a 70 ft jib with the dumping radius being 60 feet. It could slew in a full circle and was fitted with a 1½ cubic yard bucket which could excavate 400 to 500 cubic yards per day. Whitakers built two long jib machines in 1900, the one pictured and another which was sold into Oxfordshire. Both machines were still hard at work in 1927.

17 This rare 1905 drawing shows a new watermain pipe being laid in a trench prepared by the innovative Jubbs mechanical steam trencher. The eight man crew employed to operate the machine and tidy up the earthworks would have excavated some 20 to 30 yards of trench each day. The machine would then have been used to lower the pipe into the trench and to backfill. The Jubbs Patent Trencher is very similar to the modern crawler excavator in that the one pictured could dig below ground level and back towards the machine. No records exist of how many were produced and none have survived.

18 The skimmer scoop excavator attachment was an American invention which was imported and pioneered in the UK by Priestman Bros. of Hull in the 1920s. It was later taken up by most UK excavator manufacturers for fitting to excavators instead of the standard boom. The Ruston No. 4 excavator shown here in this 1926 photo, believed to have been taken in London, is hard at work skimming a stoned road surface, which will later be resurfaced with tarmacadam. The skimmer excavator was ideally suited to take only a shallow and level dig. The bucket was fitted with a trap door to the rear and was discharged by the operator pulling a rope when the boom was fully raised.

19 In 1881/2 Sir W.G. Armstrong & Co undertook the design and manufacture of the world's first hydraulic excavator, based on the lines of the Ruston tower crane. Several machines were built by Armstrong, most of which were sold to Lucas & Aired for use in the construction of the Alexandra Dock in Hull during 1882–4. One is shown here. The hoisting or digging cycle was operated by an inclined hydraulic cylinder positioned to the rear of the machine. It was of 14 inch stroke, and operating pressure was 750 PSI. The Armstrong Hydraulic Steam Excavator was reported to be very simple and reliable in work, but for some unknown reason, further development in hydraulic steam excavators did not follow for 70 years.

20 In this picture we see a Ruston railroad-type steam shovel excavating a railway cutting and loading into wagons. This machine runs on rails, and note the front two outriggers that steady the machine whilst in work.

21 Large machinery would normally have to be stripped down before it could be moved from one site to another, and this picture taken in the mid 1920s shows a 50 ton excavator loaded on to a four-wheeled Fruehauf road truck trailer. It is pulled along by two lorries, and the dragline bucket is being carried in the second lorry.

22 This steam-driven Ruston dragline is excavating sand from below water level which it dumps into the timber hopper to the left of the machine. This funnels the sand into smaller railway wagons for transport. Note the caterpillar tracks. The cabin has corrugated iron only on the top half. This was to protect the engine and gears, and full cladding cost extra.

23 Owned and operated by Concrete Aggregates of Chiswick, London, this Ruston excavator is fitted with a clam shell bucket. Also known as a 'grabbing crane', the machine excavates the sand direct from the pit, placing it in the wooden hopper to load small railway hoppers. Note the electric lights fitted to the cab and boom to allow winter and night time working.

24 A 1927 Ruston No.4 dragline excavator dredging a Fenland waterway in Lincolnshire. The heavy machine is lifting a timber mat from the rear to re-lay at the front so that it can drive on it and thereby avoid getting stuck in the mud.

25 The same Ruston dragline excavator shown a few yards further along the bankside atop the timber mats.

26 A small 1/2 cubic yard shovel powered by a petrol paraffin engine on test at a Ruston factory. A wooden cab would be added to this to complete the machine.

27 This Ruston No. 4 dragline excavator of 1927 vintage is seen here working in the Essex/Hertfordshire area in 1953. The machine is clearing out a silted-up drainage ditch with a dragline bucket. Considering the machine had been working for over 25 years, it does not seem to be in bad shape.

28&29 Photographed in 1928, this Ruston No. 4 dragline excavator has just finished cleaning out a drainage ditch. Note the clean sides of the excavation and the bleakness of the area.

30 Photographs 30-40 were taken between 1921 and 1924 and feature the Ruston No. 6 and Ruston No. 15 excavators. This first picture shows a No. 6 face shovel excavator excavating for foundations and loading a steam lorry west of London in 1921. Note the unusual road wheels fitted to the undercarriage of this excavator, which allowed the machine to travel on the highway under its own power or to be towed along behind a tractor or lorry.

31 This Ruston No. 6 face shovel excavator is pictured in 1922 on the site of the Aldershot Barracks in Hampshire, England. The machine, which is in the ownership of Frank Harris Bros. Ltd of Guildford, Surrey, is removing a hill prior to construction of the new barracks and camp. Note the Fordson industrial tractor being used to tow the trailer and three hoppers loaded with spoil.

32 This unusual and somewhat amusing picture shows a Ruston No. 6 face shovel excavator travelling through Chipping Sodbury High Street in 1922. The machine had finished work on one site and was being driven 20 miles to the next one. Just try doing this in the mid-1990s! The size of the machine is clearly shown by the onlookers and the three-storey shops and houses in the background. Note the lack of other road traffic.

33 A Ruston No. 6 dragline excavator travelling under its own power from one site to another is seen here crossing a bridge which would have been designed to take horse and carts. The machine would almost certainly have caused some structural damage to the bridge. The photograph was taken in the Fens during 1923.

34 This Ruston No. 6 face shovel is owned by the Tuberculous Ex-Service Men's Society Ltd. The machine is pictured in 1922 widening the York Road out of Leeds. Note the fleet of 5 steam wagons which the machine could easily fill on turn-round. An additional wagon would have been required to carry water and fuel to the excavator. The 150 gallon water tank for the steam engine required filling every 3 to 5 hours.

35 This Ruston No. 6 steam crane navvy excavator pictured without the corrugated iron cabin, clearly shows the steam engine boiler and operator's position. Note this machine is fitted with long bucket excavator arms which enable it to excavate to a depth of 16 ft below ground level.

36 A No. 6 dragline excavator photographed in March 1924 excavating a 16 ft deep trench in which a 4 ft wide water main was laid. The machine is working for W. Tawse of Aberdeen, Scotland, who wrote to Ruston's in its praise, saying that they had to stand the machine idle as the pipe makers could not get the pipe to the site as fast as the machine could lay it.

37 A Ruston No. 10 standard excavator owned by contractors Sir Robert McAlpine & Sons, who ran a fleet of almost a dozen similar machines. The No. 10 is shown working on a new road contract near Southampton for the Hampshire County Council in 1924.

38 & 39 These two photographs show a Ruston No. 10 Universal excavator, owned by Messrs. Pease & Partners, excavating slag from a dump and loading it into barges. The slag was transported to the docks via the railway. The second picture shows a close-up of the excavator and its driver. Note the fully enclosed cabin.

40 A Ruston No. 10 dragline excavator pictured here in 1925 on a contract to build retaining banks to hold dredging from the Manchester Ship Canal. The machine was owned by Edmund Nuttall Sons & Co. Ltd, a firm which is still trading today.

41 This drawing shows a 1925 Standard B Erie shovel loading into a cart. Note the flat wheels on the excavator, which was manufactured in the United States.

42 This 1924 photograph taken at the Oliver Iron Ore Mine in the USA shows a Bucyrus-Erie BE375 face shovel excavating deep in the mine. The huge size of the excavator can be judged from the height of the man standing at the left front wheel. Note the steam train pushing the loaded hoppers away in the top right of the picture.

43 Seen here in North America in 1928, a steam-powered Bucyrus 50B shovel is excavating the rock and loading it into the small lorry to the right. Two more scoops of the machine and the lorry would be overloaded. The coal bunker to supply the engine and the shovel is clearly visible.

44 This early model wooden-cabbed Priestman Cub of 1941 vintage is fitted with a drag shovel excavator and tile drain bucket. The machine is excavating to a depth of 18 inches across the field to enable tile drains to be laid. This work was carried out by hand by farmworkers before the arrival of mechanised excavators.

45 Another 1941 Priestman Cub excavator, fitted this time with a standard width back actor bucket. This machine, photographed in the late '40s in the ownership of Civil Engineers Fitzpatrick & Son, is excavating for a new sewer to a depth of about 6 ft.

46 A 1939 Priestman Cub clam shovel excavator working in Lincolnshire during the Second World War. The machine is fitted with a Priestman-designed level cut clam shell bucket that is clearing a silted-up drain.

47 & 48 A 1940 Priestman Cub with level cut clam shell excavator bucket. This machine has been mounted on a barge for dredging silted-up waterways. The mechanism on the level cut clam shell bucket is clearly visible.

49 Photographed in the 1950s, this metal-cabbed Priestman Wolf MKV face shovel excavator is excavating below ground level and loading into the tipper lorry. Despite its size, the machine is clearly capable of working in a confined space, as the work of tamping new-laid concrete can be seen taking place.

50 Fitted with a standard dragline excavator, this Ransomes & Rapier machine excavating a new ditch is standing directly in the line of pull. Photographed towards the end of the Second World War, the machine is fitted with a 30 ft crane jib.

51 The small Ransomes & Rapier 410 excavator is working in a chalk pit with a face shovel attachment, loading chalk directly into a hopper to be carried away on an endless belt for use in cement manufacture. Note the excellent 180 degree visibility of the working area from the operating position. Only a few of these Rapier 410s survive.

52 Viewed from the driver's side is another Rapier 410 excavator. This time the machine is fitted with a lattice type crane jib and dragline bucket. The wide driver's door is clearly visible and shows the cramped operator's position.

53 Another view of the 410, this time filling in a pond. A sliding door on the machine cabin gives access to the engine and drums to allow maintenance and refuelling.

54 This MK2 Rapier 410 excavator is owned by Civil Engineers Dowsett and carries fleet No. E15, which refers to excavator No. 15. The machine is fitted with a drag shovel, which is clearing to allow a new sewer to be laid. The photograph dates from 1953.

55 Photographed in the Norwich area, this Rapier 410 drag shovel fitted with skimmer bucket is stripping top soil from the site of a new gravel pit. The lorry in the background is tipping at the edge of the field.

56 Another Ransomes & Rapier 410 excavator with back actor drag shovel attachment excavating a trench for a new sewer on the site of a post-war housing estate. The new road has been marked out and excavated by a skimmer excavator similar to the machine in photograph 55.

57 This shot of the Rapier 410 skimmer-type excavator gives us a closer view of the skimmer bucket and sliding mechanism. Note the stop plate fitted to the lower part of the boom, which is used to fit the dredge shovel and bucket to allow fast interchange of equipment.

58 Another Rapier 410 skimmer-type excavator removing a bank. This photo was taken in the early 1950s.

59 Photographed at Frank Perkins, Peterborough is this Ransomes & Rapier 410 dragline excavator of 1950 vintage. The machine was on display at the entrance to the factory to show the variety of machines and applications of Perkins diesel engines.

60 This Ransomes & Rapier 414 drag shovel rope excavator is working on the roadside excavating a trench for a water main and loading directly into a lorry. Note the sliding doors giving access to the engine room and winding drums.

61 Drag shovel excavator digging a drainage ditch beside a newly constructed 'A' road in the 1950s. Excess spoil is loaded directly into the tipper lorry alongside.

62 Owned by Kelley of Peterborough, this Ransomes & Rapier 414 drag shovel is excavating a trench in the road to facilitate the laying of a new water main. Pictured here at the start of a day, the operator is just completing his greasing up. Note the traffic lights that have only two colours, also the lack of any road traffic.

63 Very popular for military use during the 1950s was the Blaw-Knox excavator. This BK50 drag shovel in civilian use is excavating footings for a new housing estate.

64 Another shot of the BK50. This time the machine is a prototype and fitted with a face shovel. The machine was powered by a Perkins diesel engine.

65 This Blaw-Knox BK50 drag shovel excavator is excavating footings and drainage work on a new housing estate. Note the man clearing away excess spoil and tidying up the trenches.

66 This front view of the Blaw-Knox BK50 excavator shows the width and ground clearance of the tracks and undercarriage of the machine.

67-69 These photographs show the BK50 excavator at work on the same housing site as before, excavating the footings. Although heavy, cumbersome and slow when compared with a modern hydraulic excavator, it was nevertheless much faster than by hand.

70 The Smith 210 excavator manufactured in Yorkshire, England by Thomas Smith & Son of Rodley. In both size and appearance this machine was designed to be an exact image of the very popular Ruston-Bucyrus 10RB excavator. It is seen here fitted with a face shovel working in a sand and gravel pit. Note the wheelbarrow with a metal wheel.

71 Smith-Rodley 210 excavator fitted with standard-type drag shovel excavator. Because of the light weight of the 210, it is pulled towards work when excavating in line. This operator is working with the tracks sideways towards the pull, thus ensuring that the machine stays put.

72 Working in a quarry, this Smith-Rodley model Super 10 face shovel excavator is loading into a hopper, feeding a continuous belt to remove the excavated material.

73 This Smith 21 face shovel excavator is owned and operated by Welford Gravels Ltd. Note the work lights fitted to the machine to allow for winter use.

74 A 1930s Ruston-Bucyrus 10RB skimmer-type excavator owned by Bernard Pumfrey of Gainsborough. Note the flat platform at the rear of the machine which has been added to assist the operator in refuelling and starting the engine.

75 This 1940s 10RB face shovel is pictured here excavating soil which is loaded into the small tipper lorry. This is an original Ruston publicity photograph which has been touched up by their artists.

76 Pictured in a quarry setting, this 1940s 10RB face shovel excavator is loading into the Muir-Hill dumper which can be seen behind the machine. Although small, the little 10RB was ideally matched to Muir-Hill and Chaseside dump trucks.

77 This 10RB excavator is fitted with a clam shell bucket and is being used for moving stone chippings from the stock pile into the delivery lorries. The 10RB excavator and the lorry are owned and operated by the Lea Valley Sand and Ballast Co.

78 This Ruston-Bucyrus publicity photograph shows a continental-cabbed 10RB dragline excavator. Note the tapered sloping cab roof edges which were designed for rail transport through British and continental European railway tunnels, hence the name continental cab.

79 The small Ruston-Bucyrus 10RB excavator was well suited for handling work in quarries, sand and clay pits, as this example clearly shows. The 10RB is fitted with a face shovel and is loading material into a hopper positioned over a conveyor belt.

80 This photograph shows a 1940 Ruston-Bucyrus 10RB dragline excavator with a 40 ft lattice boom and clam shell bucket. The machine is at work loading railway wagons from a stockpile.

81 This photo shows a 10RB with drag shovel/back actor excavator attachment owned by R.C. Hammet and working in Kent. The banksman is taking a risk by standing directly in the path of the working excavator, as it would only take a steel wire to break when slewing to cause an accident. The drag shovel attachment used the same boom as the skimmer bucket.

82 Transporting the 10RB excavators from site to site was not as difficult as were movements experienced with some of the larger excavators, which needed to be partially dismantled in order to move to the next site. This new Ruston-Bucyrus 10RB skimmer excavator is shown outside the Lincoln factory of Ruston's awaiting delivery to the contractor. It is being moved on a 6 axle ridged body lorry belonging to Saltergate Garage of Lincoln, who were used by Ruston-Bucyrus to deliver their smaller excavators.

83 This picture shows a late 1930s Ruston-Bucyrus 10RB face shovel excavator which has just left Ruston's Lincoln factory for delivery to the customer. The machine is loaded onto an early articulated low loader lorry in the ownership of Saltergate Garage. The extra length of the trailer allowed the machine to be transported with the face shovel excavator equipment still attached, thus saving time in dismantling and re-assembling the machine before and after transport.

84 Photographed in Australia, this 10RB face shovel excavator is working near Canberra during the early 1940s. It is being used as a crane moving granite blocks. The machine was built in Australia by Ruston Hornsby, but was identical to the UK built machines.

85 This UK built Ruston-Bucyrus 10RB face shovel excavator is pictured in 1946 at work in Rouen, France digging a tunnel. The small size of the 10RB made it ideal for working in tight corners, but the picture shows that there was very little room for mistakes.

86 Another 10RB face shovel at work in Australia excavating a basement for a new department store in Melbourne. Because the cab has been removed, the machine looks considerably smaller than usual.

87 A Ruston-Bucyrus 10RB skimmer photographed in the late 1930s, excavating a roadway into a new housing estate in the London area. The excavator is loading soil straight into the waiting lorry.

88 Tarslag engineers based in Wolverhampton were the first company to purchase a 10RB excavator in 1934. This machine of slightly later vintage is excavating foundations for a new estate road in London just before the start of the Second World War in 1939. The 10RB is fitted with a skimmer-type excavator bucket.

89-91 The drag shovel excavator, also known as a back actor or back hoe excavator, was usually seen on civil engineering and house building works. These three photographs show the Ruston-Bucyrus 10RB drag shovel excavator at work in 1945/46 excavating trenches for drains and pipelaying on a new housing estate.

92 & 93 The basic dragline or clam shell excavator could easily be converted into a crane by removing the bucket and replacing it with a hook. The picture above shows the Ruston 10RB with a 28 foot channel jib being used as a crane at the French coal mines in the Sarre/Moselle. Note that the 10RB is electrically powered as opposed to the more usual diesel engine version. The picture to the right shows a 10RB fitted with a 32 foot lattice-type jib working in England for contractors Joseph Parks & Son.

94-97 Pictured on these two pages working for Blackford's of Calne, Wiltshire in the 1940s is a Smith-Rodley dragline excavator. The machine is rigged as a crane and is unloading and placing materials before repairs and dredging take place. Note the large crowd of onlookers the machine draws. The machine is diesel powered and carries the registration number AXN623.

98 & 99 This Smith-Rodley dragline excavator is pictured working in Calne, Wiltshire, on bridge and river works reinstatement.

100 A Bucyrus-Erie BE375 dragline excavator stripping phosphate in Plant City, Florida, USA.

101 One of the first Priestman Bros. trailed-type grab excavators. Operated by two men, one driving the tractor and one operating the excavator, this 1920 machine was simple and basic in its design and operation, but set the pace for Priestman excavators for the next 20 to 30 years.

102 Keystone model 4-26 skimmer-type excavator. Manufactured in the United States by the Keystone Driller Co. of Beaver Falls, Pennsylvania, the machine was powered by a combustion engine. A set of caterpillar-type crawler tracks were chain driven to provide forward and reverse travel, and two single iron wheels were used for steering. The machine was ideal for clearing a few inches off the road surface prior to black-top being laid. The machine is pictured outside the Keystone factory.

103 Priestman Bros. of Hull produced this self-propelled skimmer excavator in 1923. It was based on the American Keystone excavator and the similarities are clear. None of these machines are known to have survived, and it is uncertain how many were built.

104 This series of photographs was produced by Priestman's publicity department in April 1933. It shows the River Douglas near Bank Bridge, Tarleton, where the Lancashire County Council drainage department assembled a 15 ton sectional steel pontoon. It was designed to carry a Priestman diesel dredging grab which was used to clear clay silt and sand bars which were blocking the mouth of the river. Priestman received a number of orders for similar pontoons following the success of this work.

105 & 106 To achieve complete dredging and clearing out of waterways and rivers, it was not always possible to work the excavator from river side, and so a number of excavators were permanently mounted on barges. These two photographs show steam and diesel barge-mounted Priestman excavators at work.

107 **& *108*** These photographs taken in 1932 are of Priestman No. 15 Universal 1/2 cubic yard long-radius dragline excavators. The jibs were 40 ft long. The machine to the right is working on the river Ouse in Yorkshire. The machine below is owned by the River Hull Catchment Board and is deepening a tributary on the River Hull. On this sort of work the machines could complete 44 lineal yards per day, which represented an hourly silt excavation rate of 25 cubic yards per hour.

109 This Priestman excavator is dredging out a river at low tide and tipping the silt to the rear of the machine. Note the wooden sleepers that the machine is working from and the night working lights fitted to the front of the machine. The excavator is powered by a 75 brake horsepower McLaren diesel engine. It has a 60 ft long lattice jib with a ¾ cubic yard dragline bucket. The machine weighs in at 32 tons and was supplied to the River Ouse Catchment Board in Yorkshire. The photograph dates from about 1935.

110 Shown working at low tide, this Priestman Panther excavator is owned by the Somerset Rivers Catchment Board. It has a 70 ft long jib with a ¼ cubic yard dragline and weighs 29 tons. The machine was powered by a 65 brake horsepower McLaren diesel engine.

111 This is a Priestman Tiger dragline fitted with a specialist levelling blade which was used to drag heaps of tipped spoil which required levelling off. The chimneystack in the picture is not part of the machine and is in fact a building positioned a few hundred yards away.

112 This photograph shows the distance that a dragline excavator can operate from. The machine is clearing about 40ft of silted river bank, which is excavated and dumped to the rear of the machine. Note the beautifully clean, finished work which can be achieved by an experienced driver.

113 This Priestman Panther dragline excavator is seen here operating in North Lincolnshire. It is fitted with a heavy duty 3/8 cubic yard excavator bucket and has a 40 ft long open lattice jib. The machine was photographed just after the end of the Second World War.

114 This Priestman 1/2 cubic yard diesel-powered dragline excavator is fitted with a 40 ft lattice-type jib. It was supplied in September 1932 to the River Ouse Catchment Board in Yorkshire and is shown here at work on the River Derwent.

115 This Priestman Bros. publicity photograph shows clearly the 115 ft slewing range of the excavator for digging and dumping works. This machine is owned by the River Trent Catchment Board and was supplied to them in 1934.

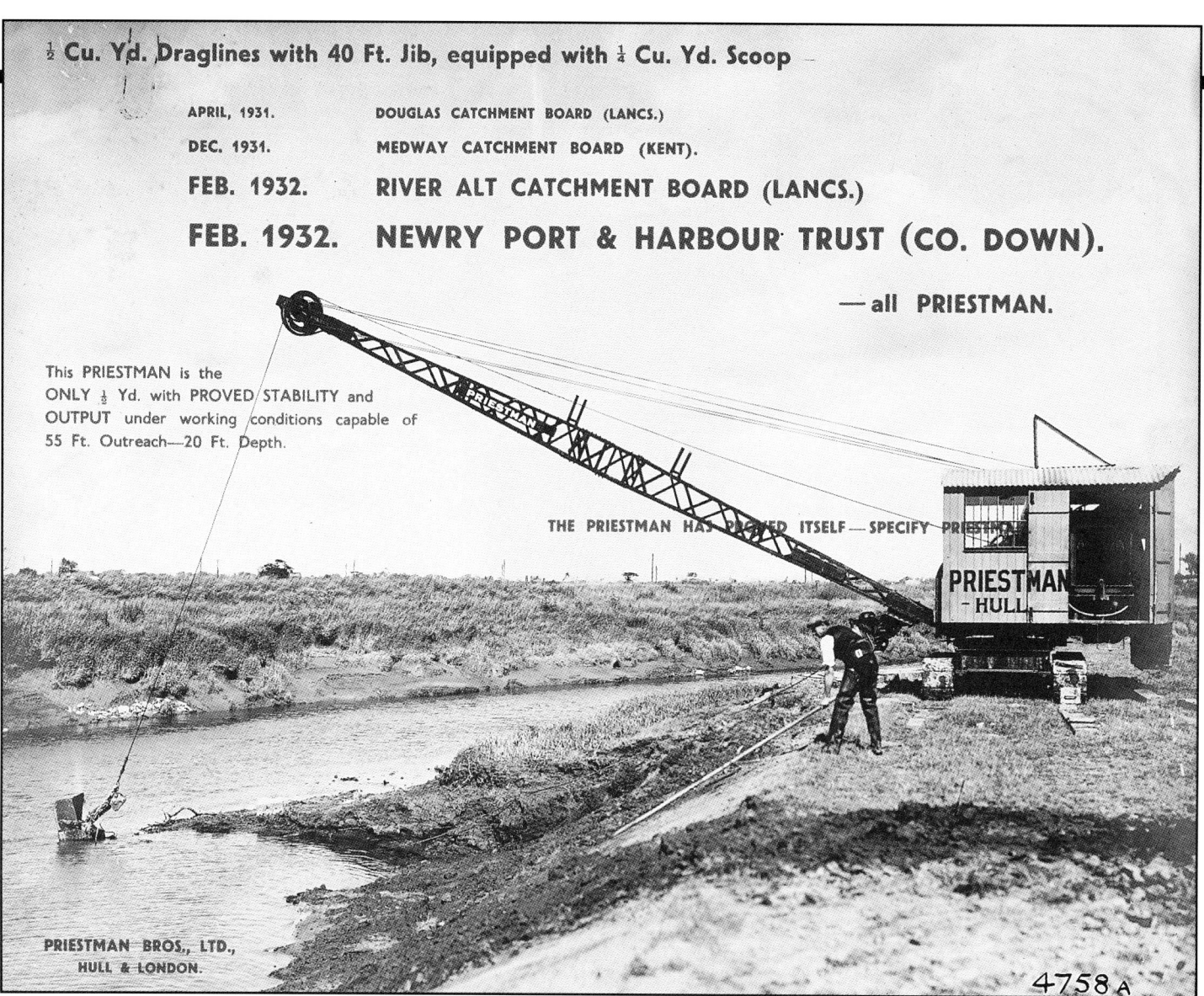

116 The sales department of Priestman Bros. set great store in the high number of River Catchment Boards that purchased their excavators. They were always quick to see the publicity value of orders for new machines, and this 1932 photograph shows the new Tiger excavator supplied to the River Alt Catchment Board in Lancashire.

117 This shows a 1935 Priestman Panther No. 12 diesel-powered excavator which was supplied new to the Lancashire County Council. The 12 referred to the weight of 12 tons.

118 Another Priestman Panther dragline excavator in East Anglia in the mid 1930s.

119 Another shot of the Priestman Panther 12 supplied to Lancashire County Council. The machine has a 55 ft lattice-type jib. Note the wooden timbers upon which the excavator rests, also the metal block under the track to alleviate any rocking of the machine whilst in work.

120 One of the most popular excavators of the 1930s and '40s was the small Priestman Cub. The one pictured is in the ownership of the River Severn Catchment Board based in Worcester. The machine is fitted with a 32 ft lattice-type jib and 1/2 cubic yard dragline excavator bucket. The main competition for the Cub was the Ruston-Bucyrus 10RB.

121 The two Priestman excavators pictured here just before the Second World War are widening a drain in the Lincolnshire area.

122 Photographed at the Priestman Bros. Hull factory in 1944, a line-up of excavators awaiting dispatch to customers. The first four machines are dragline excavators, the first two of which are fitted with standard channel jibs. The second machine is fitted with a side dragline arm to facilitate the cleaning out of ditches. The third and fourth have lattice-type jibs. The fifth machine, less cab, is fitted with a face shovel, and the sixth machine has an unusual skimmer dozer blade.

123 This set of photographs was taken at the Priestman Bros. Training School in East Yorkshire during March 1941. The men pictured were being taught how to set out the work and maintain and operate a Priestman Cub excavator. These courses were run by Priestman from the 1920s onwards.

124 Four more photographs from the Priestman Training School. This time the trainees are being taught how to winch out a Priestman Cub excavator which has become stuck in a ditch. They are using a Trewhella winch and earth anchor. Note the Land Army girls who are also being trained.

125 Pictured here on the River Dee in the early 1940s are two Priestman grabbing crane excavators. The machine working from the bank is a wooden-cabbed Priestman Cub, which is dredging the river and tipping the spoil over the wooden spikes in order to strengthen the river bank. Meanwhile, working deeper in the river is an older machine mounted on a pontoon and loading directly into a barge.

126 Another barge-mounted Priestman grabbing excavator which is dredging the river bank and dumping spoil directly into the barge berthed at the side.

127 Working within the territories of the River Severn Catchment Board is this pontoon-mounted Priestman Panther dragline excavator which is dredging underneath the bridge whilst waiting for another barge to come alongside.

128 This Priestman grabbing crane excavator has been permanently fitted in this ship to enable deeper waters, mud flats and sand banks to be dredged.

129 This photograph shows a rare Priestman Cub clam shell excavator fitted with rubber tyres and front-wheel steering in lieu of the usual caterpillar-type tracks. The machine was built in 1937 for rehandling sand and ballast. Its cost new was £850, and the machine shown was owned by Lavenders. Note the diesel engine, radiator on the side of the cab, and the lattice-type jib.

130 This 1950 Priestman Cub excavator is of an unusual high-cab design with a rear-mounted engine. It is fitted with a standard channel boom and a side dragline attachment which enabled it to clear out a ditch to the side of the machine.

131 Supplied to the Yorkshire Land and Warping Co. during the Second World War, this Priestman Cub dragline excavator is shown deepening and widening a drain. The machine has a 40 ft outreach.

132 & 133 This 1940s Priestman Cub excavator is cleaning out a 120 ft wide lake, using a scraper scoop instead of a standard dragline bucket. This equipment will excavate only a thin slice at a time, thus ensuring that the machine does not dig through the clay base of the man-made lake to expose the porous chalk below. The machine carries registration number BAT372 and is owned by F. Ridley.

134 Another wooden-cabbed Priestman Cub, pictured clearing out a drainage dyke for the River Trent Catchment Board.

135 A Priestman publicity advert.

136 A Priestman Cub in operation 1944/6.

137 Pictured here in the West Midlands is a Priestman Wolf MK4 face shovel excavator, which is truck mounted. The machine was originally owned by the British Army's Royal Engineers based at Chatham in Kent.

138 & 139 These two photographs show the top-secret 1940 Ruston & Hornsby Nellie excavator on test at Clumber Park in Lincolnshire being watched over by Sir Winston Churchill. Churchill instructed the Director of Naval Construction to develop a giant 'mole' to enable a surprise attack on the Siegfried line. It was to cross the 5 mile wide no-man's-land under cover of darkness, creating a trench through which men and tanks could move undetected by the Germans.

Although the machine's code name was Cultivator No. 6, technicians referred to it as Nellie, from NLE for Naval Land Equipment. It weighed 120 tons, and was 77 ft long, 6$^1/_2$ ft wide and 8 ft high. In an hour it could travel a mile whilst moving 8,000 tons of soil. Churchill ordered 250 machines, but after only three were built, the course of the war had altered to the point that they were no longer needed. Fittingly enough, the 600 horsepower diesel engines specially designed for Nellie by Harry Ricardo and built by Davey, Paxman & Co. of Colchester, powered the landing craft used in the Normandy invasion.

140 & 141 This Ruston-Bucyrus 19RB face shovel excavator sporting the army number 10 BY20 is featured at work in the UK, loading a Fordson Major-based dump truck. Note the face shovel boom which has details of the army driving requirements stencilled on the side. The 19RB was built in 1947, and the photos date from 1955.

142 This 1944/5 Ruston-Bucyrus 19RB face shovel excavator is loaded up awaiting its turn in the 1946 Victory parade held in London. The boom height and length allowed the machine to be transported under bridges on the main road network. These machines almost certainly saw active service in Europe and would have been an essential piece of equipment for the army to take to the battlefront, vital for clearing a way through the rubble and for repairing roads and airfields.

143 A fine display of face shovel excavators busy moving sand and earth to widen a roadway, possibly to allow troop movements or military vehicles to pass by. The excavators pictured are the British Ruston-Bucyrus 19RB and the North American designed NCK304.

144 Under the supervision of the British Army's Roads and Airfields section, three 19RB draglines, together with an American lend-lease Caterpillar D8 bulldozer. A crawler tractor and a tracklaying dumper truck can also be seen to the rear left of the photograph.

145 Photographed at Nijmegen in Holland during 1945, this large grabbing crane excavator is being pressed back into service by the British Army to assist with clearing a bridge following the withdrawal of German troops. The excavator is barge mounted and steam powered.

146 This 1943 North American Bucyrus 10B drag shovel excavator is excavating a trench and loading the spoil into the awaiting dumper. The machine, which was brought over from the USA under lend-lease, has been named Valerie after the operator's sweetheart. E22 is the fleet number. This machine was virtually identical to the Ruston-Bucyrus 10RB excavator except for the make of engine and the cab shape.

147 For comparison with the Bucyrus 10B excavator, photographs 147 and 148 of British-built Ruston-Bucyrus 10RB drag shovel and skimmer excavators, in which the different cab can easily be seen. The excavator is digging a trench for installation of a new water main in the north of England during 1953. The same boom is used on the drag shovel and for the skimmer bucket, thus allowing quick change of excavator attachments.

148 Photographed in 1943, this Ruston-Bucyrus 10RB skimmer excavator is being operated by a sapper in the Royal Engineers. The machine has a 3/8 cubic yard bucket and will out-perform a tractor-type loading shovel on short-cycle loading work. The 10RB has a continental cab.

149 In post-war Britain many large contractors needed excavators to help with the rebuilding work being carried out. Few new machines were available but there was a plentiful supply of virtually new and sometimes unused ex-military excavators for sale. This photograph from about 1950/51 shows one such machine, a Ruston-Bucyrus 21RB face shovel excavator in the ownership of George Wimpey Ltd. Note the dragline excavator working in the distance.

150 Not all excavators were of the smaller size of those featured in this book. Some, like the gigantic face shovel excavator pictured, were extremely large. The true size of this monster can be seen from the driver sitting in the smaller excavator in front, which was the size of the Ruston-Bucyrus 21RB excavator featured in photograph 149. Clearly this machine could not have been moved from one site to another by road, as it would have had to be stripped down and transported in sections. An excavator of this size would usually spend its working life on a single site in an open-cast coal mine or quarry.

151 This Ruston-Bucyrus 50RB dragline excavator is in the ownership of Sir Robert McAlpine of London and is pictured here loading a lorry with earth. This operation required a very skilled operator, as the bucket would swing around and could easily crash into the lorry, which would certainly be knocked over.

152 & ***153*** The Blaw-Knox BK50 excavator shown has been truck mounted in the Royal Engineers' own workshops to provide them with an excavator which can travel long distances at reasonable speed. The excavator operator can also move the truck forwards and backwards. The range of excavators and attachments are part of the equipment used during the late 1940s and early 1950s by the Royal Engineers School of Military Engineering at Chatham in Kent.

154 The Insley Manufacturing Co. of Indiana built its first excavator during 1924. Many of its model K12 excavators were imported into the UK in the years following the Second World War. Photographs 154-170, taken in the USA, date from the mid 1940s. The first shows an Insley K12 face shovel excavator loading a truck on a highway site in North America.

155 & 156 The Insley K12 excavator was similar in size to the UK built Ruston-Bucyrus 10RB. In the late 1940s Insley were selling the K12 face shovel with an electric bucket trip.

157 & 158 Insley excavators shown at work on a new housing estate during 1946. Note the rear weight fitted to the machines to allow the 1/2 cubic yard bucket to be filled up to the maximum.

159 & ***160*** At work excavating a trench for a 4 inch gas main is this 1940s Insley model K12 drag shovel excavator.

161 Although the back hoe was lighter than, for example, the face shovel, it was still able to excavate in rock. The machine shown is excavating a trench for a new water main on a housing estate.

162 Rear view of the Insley K12 back hoe excavator, which is digging a trench 5ft deep in which to lay a new water main. Note the heavy counter weight attached to the rear of the excavator.

163 The Insley back hoe excavator shown in this photograph was recorded as having excavated a run of 900 to 1200 ft to a depth of 5ft on a contract in 1946.

164 This picture shows the crane-type excavator attachment available for the standard excavator, which is fitted with a 30ft boom to which extensions of 5ft and 10ft could be added to increase the reach. The machine is fitted with a clam shell bucket working on the site of a new road.

165 This machine is cleaning up the sides of a drainage ditch with a standard $1/2$ cubic yard dragline bucket.

166 The excavator clam shell bucket could be quickly detached and replaced with the crane hook to allow for pipes, piles and sheets of steel to be moved on site without the need to have a second costly machine waiting. The picture shows an Insley K12 clam shell excavator digging a trench, after which it will lower the new pipes and then backfill the excavation.

167-169 These three examples of the Insley K12 excavator are of the more unusual lorry mounted excavators. Many of these machines were operated by the British Army, but few were sold for use in the UK construction industry, which seemed to prefer the crawler tracklaying version. The first photo is of a crane-type machine fitted with a 'skull cracker' implement, which was dropped at speed and would compact or crack the road surface. The second photo is of a face shovel excavator and the last shows a clam shell bucket excavator.

170 An experienced operator would be needed to carry out this task of excavating with a dragline excavator and then loading directly into a waiting lorry. One slight miss and the heavy dragline bucket would smash the lorry.

171-174 Insley offered a specially built trailer on which to move their K12 excavator from site to site. It was suitable for towing behind a small lorry. The front turntable axle was removed, the machine was driven onto the trailer and the excavator boom was attached to the trailer and used to raise it, so as to re-position the front axle. The trailer was then hitched up to the lorry and was ready for transport. The whole operation took only 10 minutes.

175 This unusual excavator, maker unknown, is shown working in East Anglia in 1951. It is a tracklaying chain and bucket-type which worked the same way as a chainsaw blade. The buckets tipped the spoil into a hopper and the material was moved over to the other side of the ditch by means of a conveyor belt.

176 An early 1950s Ruston-Bucyrus 38RB face shovel excavator. This machine is working in a quarry excavating from a rock face and loading into a dump truck. The 38RB was powered by the 132 horsepower Ruston 6 VPHN diesel engine.

177 A Ruston-Bucyrus 54RB 2½ cubic yard dragline excavator. This machine is fitted with a 60 ft boom with a 58 ft radius. Optionally it could handle a massive 100 ft boom with a 95ft radius and a 1 cubic yard dragline bucket. The machine pictured weighs 69 tons.

178 A Ruston-Bucyrus 100RB face shovel working in a quarry. This 135 ton monster has a 3½ cubic yard front bucket. The machine is electric powered by a Ward Leonard electric induction motor of 200 horsepower.

Index

Italic numbers refer to caption numbers, bold to page numbers in the introduction.

A
Armstrong Hydraulic Steam Excavator, *19*

B
Barclay Steam Shovel, **9**, *10*
Blaw-Knox BK50, *63, 64, 65, 66, 67, 68, 69, 152, 153*
Bucyrus 50B, *21, 43*
Bucyrus-Erie BE375, *42, 100*
10B, *146*

C
Caterpillar D8, *144*
Chaplin Steam Navvy, **9**, *9*

D
dragline excavator, *112*,
Dunbar Steam Crane Navvy **see** *Ruston-Dunbar Steam Crane Navvy*

E
Erie Standard B shovel, *41*

F
face shovel excavator, *150*
Fordson industrial tractor, *31*
Fowler/Waite scoop, *2*
Fruehauf road truck trailer, *21*

G
grabbing crane excavator, *145*
Grimshaw Steam Excavator, *7*

I
Insley K12, *154, 155, 156, 157, 158, 159, 160, 161, 162, 163, 164, 165, 166, 167, 168, 169, 170, 171, 172, 173, 174*

J
JCB JS200LC, *1*
JCB MK1 Excavator, **10**
Jubbs Mechanical Steam Trencher, *17*
Jubilee excavator, *11*

K
Keystone model 4-26, *102*

L
Lima excavators, **11**
Lorain excavators, **11**

M
Muir-Hill dumper, *76*

N
NCK304, *143*

O
Osgood dragline excavator, **10**
Otis Steam Shovel, **7-8, 9**, *3, 4*

P
Priestman
  No.1, **10**
  No.5, **10**
  No.15 Universal, *107, 108, 114, 115*
  barge-mounted excavators, *105, 106, 126, 127*
  Cub, **10-11**, *44, 45, 46, 47, 48, 120, 123, 124, 125, 129, 130, 131, 132, 133, 134, 135, 136*
  dragline excavators, *109, 114, 115, 122, 127*
  dredging grab, *104*
  grabbing crane excavator, *125, 126, 128*
  Panther, *110, 113, 117, 118, 119, 121, 122, 127, 128*
  pontoons, *104, 125*
  self-propelled skimmer excavators, *103*
  Tiger, *111, 116*
  trailer-type grab excavator, *101*
  Wolf, **11,** *49, 137*

R
Ransomes & Rapier
  410, **11,** *51, 52, 53, 54, 55, 56, 57, 58, 59*
  412, **11**
  414, *60, 62*
  dragline excavator, *50*
Ruston
  No. 4, *18, 24, 25, 26, 27, 28, 29*
  No. 6, *30, 31, 32, 33, 34, 35, 36*
  No. 10, *22, 23, 37, 38, 39, 40*
  railroad-type steam shovel, *20*
  steam dragline excavator, *22, 23*
Ruston-Bucyrus
  10RB, **11,** *74, 75, 76, 77, 78, 79, 80, 81, 82, 83, 84, 85, 86, 87, 88, 89, 90, 91, 92, 93, 147, 148*
  19RB, **11,** *140, 141, 142, 143, 144*
  21RB, *149*
  38RB, *176*
  50RB, *151*
  54RB, *177*
  100RB, *178*
Ruston-Dunbar Steam Crane Navvy, **9**, *5, 6, 7, 8*
Ruston & Hornsby 'Nellie' excavator, *138, 139*

S
Smith
  14, **11**
  21, **11,** *73*
Smith-Rodley
  210, *70, 71*
  dragline excavator, *94, 95, 96, 97, 98, 99*
  Super 10, *72*
Standard B Erie Shovel, *41*

T
Trewhella winch, *124*

W
Whitaker
  5 ton steam excavator, *14*
  10 ton steam excavator, **9, 11,** *12*
  long jib shovel, *16*
Whitlock Dinkum Diggers, *10*
Wilson Steam Navvy, *15*

Picture credits

Capt. (Retd.) R. T. Arnold Royal Engineers

W. Barnes, Excavating Machinery
(Ernest Benn, 1928)

Paul Clark

Contract Journal

Peter N. Grimshaw

Keith J. Haddock

Ray Hooley

JCB (Excavators) Ltd

Stephen Moate

Bernard Newman, One Hundred Years of
Good Company   (Ruston & Hornsby 1957)

Perkins of Peterborough

Priestman Bros

R B International

Rural History Centre,
University of Reading

The Vintage Plant Club Archives

Jennie Starkey

All other photographs are from the author's own collection

FARMING PRESS BOOKS & VIDEOS

If you have enjoyed this book, you may be interested in other books and videos on similar subjects.
For more information or for a free illustrated catalogue of all our publications please contact:

Farming Press
Miller Freeman Professional Ltd
2 Wharfedale Road, Ipswich IP1 4LG, United Kingdom   Telephone (01473) 241122   Fax (01473) 240501

VIDEOS

Classic Farm Machinery *Brian Bell*
  Vol I 1940-70
  Vol II 1970-95
Archive film extracts tracing the mechanisation of the chief arable operations.

Classic Tractors *Brian Bell*
Archive film extracts focusing on the development of tractors from 1945 to the present.

Farming with Steam
Shows how steam was used on the farm in threshing, ploughing and hauling and recalls the life of a traction engine driver.

Fordson, the Story of a Tractor
The Massey-Ferguson Tractor Story *Michael Williams*
John Deere Two-Cylinder Tractors
  (Volumes 1 & 2)
Videos showing in detail the machines produced by these leading companies.

BOOKS

Tractors at Work *Stuart Gibbard*
Vols I & II
Each book contains some 180 photographs spanning 1904 to the present and showing a wide range of tractors in many working situations on farms in Britain.

World Harvesters *Bill Huxley*
Over 135 photographs of equipment used in harvesting crops from all over the world.

Tractors Since 1889 *Michael Williams*
An overview in words and pictures of the main developments in farm tractors from their stationary steam engine origins to the potential for satellite navigation.

Ford Tractor Conversions: the story of Doe, Chaseside, Northrop, Muir-Hill, Matbro and Bray *Stuart Gibbard*
Detailed, profusely illustrated account of the main models and machines produced by these leading companies.

Farming Press is a division of Miller Freeman Professional Ltd which provides a wide range of media services in agriculture and allied businesses. Among the magazines published by the group are **Arable Farming**, **Dairy Farmer**, **Farming News**, **Pig Farming** and **What's New in Farming**. For a specimen copy of any of these please contact the address above.

Profitable Excavating

with a
RUSTON No. 4
Full Circle, ½ Cu. Yard
Universal Excavator

The Ruston No. 4 Full Circle, ½-cu. yard Universal Excavator is the ideal machine for small jobs especially where mobility is a consideration. Over 150 machines are at work in Great Britain alone besides others in every corner of the world.

Crude Oil—Petrol—Paraffin—Electric

As a PROFIT MAKER for Quarry Work, Road Making, Foundation Excavation, Trenches, Cleaning and Widening Rivers, Drains, etc. etc.
The Ruston No. 4 is Supreme.

Shovel—Dragline—Grab—Skimmer—Trencher
Write for Catalogue JM 5366.

Ruston & Hornsby, Ltd., Lincoln

The 1947 ½ Yard Insley Excavator

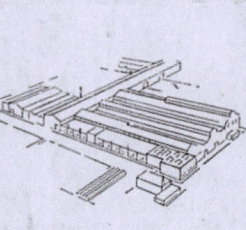

The Insley Line: Shovels, Hoes, Draglines, Clamshells and Cranes (gasoline or diesel powered; crawler or rubber mounted). . . . Concrete Handling Equipment.